Máira Beatriz Teixeira da Costa
Márcio da Silva Araújo
Gustavo Nunes Pereira

Recovering degraded areas along power transmission lines

Máira Beatriz Teixeira da Costa
Márcio da Silva Araújo
Gustavo Nunes Pereira

Recovering degraded areas along power transmission lines

A look at environmental impacts

Imprint

Any brand names and product names mentioned in this book are subject to trademark, brand or patent protection and are trademarks or registered trademarks of their respective holders. The use of brand names, product names, common names, trade names, product descriptions etc. even without a particular marking in this work is in no way to be construed to mean that such names may be regarded as unrestricted in respect of trademark and brand protection legislation and could thus be used by anyone.

Cover image: www.ingimage.com

This book is a translation from the original published under ISBN 978-613-9-62467-6.

Publisher:
Sciencia Scripts
is a trademark of
Dodo Books Indian Ocean Ltd. and OmniScriptum S.R.L publishing group

120 High Road, East Finchley, London, N2 9ED, United Kingdom
Str. Armeneasca 28/1, office 1, Chisinau MD-2012, Republic of Moldova, Europe
Printed at: see last page
ISBN: 978-620-7-72753-7

SUMMARY

Dedication...

I dedicate this work first of all to God, for being essential in my life and the author of my destiny, to my family, my partner, *teachers and* friends, for being the greatest encouragers of my dream, through the effort, affection, love, friendship and education they give me!

Thank you...

First of all, to God, for enlightening me and giving me the strength to achieve my goals with dignity and respect for others, and for allowing me to have unique experiences together with special people I have met or who were *already part of* my life.

To my mother, Delvana Quintino Teixeira, and to my family, especially my grandparents, uncles, Lilia, Elielton, Clàudio, Débora and Kâtia, to whom I owe my education, because they have always encouraged me and provided me with moments of *great* happiness, love and friendship. You are everything to me, and I can't thank you in any other way than by saying so: I love you unconditionally, thank you.

To my partner Marcelo Félix Luiz, and his family, for the love, support, affection and happiness they have given me throughout my education, and for making it possible for me to get this far.

To my advisors Màrcio da Silva Araùjo and Ismael Martins Pereira, who shared their knowledge and time not only during the preparation of this work, but also throughout my academic life.

The company Sào Simào, for the wonderful opportunity to do my internship in such a welcoming company. To its employees, especially Gustavo Nunes Pereira, who shared his knowledge and attention during my internship and who encouraged me so much.

To my colleagues and friends from the technical course and the university who have cheered me on and always helped me throughout my academic career, especially Ingrid, Jordana, Paloma, Niliane, Samara, Nayara, Franciele, Raiane, Leidimila and Jucilene. To the *teachers* and staff of the University Unit of Ipameri and Senac, for their friendship and companionship.

Finally, to all the people who have always supported me and, in a way, stayed by my side. Please know that there is a little piece of each and every one of you inside my smile.

Thank you so much!

"Let this not be the last achievement;

May dreams never *end;*

That truths are not absolute;

That knowledge is still too little;

May faith and respect prevail;

And that I don't forget my origins."

Author Unknown

SUMMARY

COSTA, Mâira Beatriz Teixeira da Costa[1] ; ARAÙJO, Màrcio da Silva[2] . RECOVERY **OF DEGRADED AREAS IN** ELECTRIC POWER TRANSMISSION LINES, Compulsory Supervised Internship Report presented to the Universidade Estadual de Goiàs, Câmpus Ipameri as part of the requirements for obtaining the degree of Forest Engineer, Câmpus Ipameri, Universidade Estadual de Goiàs, Engenharia Florestal, Ipameri, Goiàs, Brazil, 2016, 44p.

Worldwide consumption of electricity is on the rise, sustained mainly by population growth and the relentless pursuit of socio-economic development. Thus, the growth of the electricity sector in Brazil is undeniable and, with it, comes the implementation of transmission lines, fundamental parts in the transportation and distribution of electricity from hydroelectric plants to consumer centers. Despite the benefits derived from this activity, it also has a number of environmental impacts, making it essential to apply the Degraded Areas Recovery Program - PRAD, in order to comply with current legislation and minimize the impacts caused to the environment. The general aim of the work was to present a brief bibliographical review of hydroelectric power in Brazil and its distribution throughout the country via transmission lines, and also to point out the main environmental impacts arising from the implementation of these projects and thus propose mitigation measures capable of preventing and/or mitigating environmental problems. In addition, the activities carried out during a compulsory supervised internship at the company Sào Simào Montagens e Serviços Ltda are described. The main activity was to monitor a PRAD resulting from the installation of an electricity tower in the area covering the municipality of Ipameri-GO.

Keywords: Environmental impact; Transmission line; PRAD.

---

1 Student on the Forestry Engineering course at the State University of Goiás, Ipameri Campus.
2 Lecturer in Forestry Engineering at the State University of Goiás, Ipameri Campus.

1. INTRODUCTION

Worldwide energy consumption is growing, mainly due to population growth and the relentless pursuit of economic and social development (SA, 2016). This makes it necessary to install transmission lines, as they are fundamental parts of the transportation and distribution of electricity, from hydroelectric plants to consumer centers (FERREIRA, 2011).

Currently, hydroelectricity is the main source of electricity in Brazil, accounting for around 64% of electricity (EPE, 2016). However, this supply is highly dependent on rainfall, which can vary greatly from season to season, as well as being significantly unpredictable (CAMPOS, 2010).

In addition, hydroelectric power stations are distributed in river basins that are far apart and subject to different rainfall patterns. For this reason, transmission lines in Brazil make it possible for the energy produced in regions with more rainfall to be used in other regions of the country (CAMPOS, 2010).

This coordinated operation of electricity distribution via transmission lines makes it possible to mitigate the effects of fluctuations in the rainfall regime, as well as optimizing the country's hydroelectric generation capacity by up to 20% (PIRES, 2005). In view of these characteristics, the 1996/2005 Ten-Year Expansion Plan and the 2015 Plan were created, which defined the sequence of construction of generation projects and regional interconnections necessary for the country's energy supply (SANTOS, 2011).

On the other hand, the activities related to the construction of the power transmission line are responsible for major damage to the environment, as they bring with them various environmental impacts, which influence the process of degradation and, consequently, compromise the environmental balance.

The main impacts are generated by the construction of building sites, borrow pits, dump sites and the deforestation necessary to open squares, access roads, clear the right of way, move soil for the foundations, as well as the movement of heavy equipment, assembling the structures and laying the conductor cables (OTTMANN et al., 2010).

It is therefore essential to plan the production, transmission, distribution and use of energy resources in an environmentally appropriate way, in order to minimize the environmental impacts caused by these projects (SANTOS, 1997).

The scenario of the study presented in this paper is part of the construction of the LT CC

±800 kV Xingu / Estreito Transmission Line and Associated Facilities, a project needed to transport the energy generated by the Belo Monte Hydroelectric Power Plant, in the municipality of Anapu-PA, to the Estreito Substation, located in the municipality of Ibiraci-MG, from where it will distribute energy to its consumers. This work brings with it the need to recover the areas degraded by the installation of the line, in order to avoid and/or mitigate various impacts on the environment, making it possible to resume the original use or even an alternative use.

Therefore, the aim of this work was to present a brief bibliographical review of hydroelectric power in Brazil and its distribution throughout the country via transmission lines, and also to point out the main environmental impacts arising from the implementation of these projects and thus propose mitigation measures capable of preventing and/or mitigating environmental problems.

In addition, the activities carried out during the compulsory supervised internship at the company Sao Simao Montagens e Serviços Ltda are described. The main activity was to monitor a Degraded Area Recovery Program (PRAD) resulting from the installation of an electricity tower in the area covering the municipality of Ipameri-GO.

2. literature REVIEW

2.1. ELECTRICITY IN THE WORLD AND BRAZIL

In the mid-1950s, the use of electricity transformed the way nations lived and brought many benefits (SANTOS, 2011). Nowadays, the importance of electricity is becoming increasingly clear, as it influences technological modernization and the way individuals organize themselves, triggering a process of valuing this resource (MORENO and MORENO, 2001).

Population growth, together with the relentless pursuit of economic and social development, has resulted in growing energy consumption around the world (SA, 2016). According to Trigueiro (2005), consumption has increased around 100 times since the Middle Ages.

The availability of electricity, combined with its essential applications in modern society, has made it one of the most widely used forms of energy in the world, mainly due to its ease of transportation and the low rate of losses during conversion (ROMANO, 2015).

Water is the second largest source of electricity generation in the world, with a share of approximately 18%, and the five largest producers are Canada, the United States, Brazil, China and Russia (ANNEL, 2002). Also according to ANEEL (2002), the availability of this energy on Earth is approximately 50,000 TWh per year, which corresponds to around four times the amount of electricity generated in the world today.

Brazil has a vast territory and an abundance of energy resources, which allows the country to make use of a variety of predominantly renewable energy sources, such as thermal, hydroelectric, wind, solar, etc., although they are still not very significant (MENEZES, 2015).

According to data from EPE (2016), the main source of electricity in Brazil is hydroelectric, responsible for around 64% of the country's energy matrix, and it also has one of the largest hydroelectric potentials in the world, thanks to the contribution of the Amazon River and Paranà River (ABBUD and TRANCREDI, 2010).

It is therefore considered the most renewable of the world's major economies, with 75.5% of its production coming from sources such as water (hydroelectric plants). There are other alternative energy sources in the country, however, the largest investments are concentrated in the construction of hydroelectric and thermoelectric plants, with percentages of 41.3% and 55.4%, respectively (ANNEL, 2011).

The hydroelectric potential is estimated at 260 GW, but only a quarter of this is actually used to generate energy (TOYAMA et al, 2014). This is because there are many generating parks that have not yet been exploited in the North, either due to environmental obstacles, projects that are still technically and economically unfeasible or simply due to difficulties in accessing the region (MENEZES, 2015).

The country has more than 61.5 million consumer units in 99% of Brazilian municipalities, of which the vast majority are residential, which shows that the population has greater access to the electricity grid (ANNEL, 2008). However, there are regional peculiarities to this access, as the Southeast, South and Northeast are more developed than the Midwest and North (Table 1), with an accessibility percentage of 51.13%, 17.84% and 16.98% respectively (EPE, 2015a).

In this way, the sector is constantly developing and progressing, making it dependent on government policies, research projects and investments made by companies in the sector (SANTOS, 2011).

Table 1. Brazilian energy consumption by geographic region (GWh).

Region	2013	2014
North	30.209	32.364
North East	79.694	80.746
South East	240.084	243.123
South	80.393	84.819
Midwest	32.755	34.381
Brazil	**463.134**	**475.432**

Source: EPE, 2015a.

Since 2004, with the introduction of the New Electricity Sector Model known as "Light for All", with the backing of Laws No. 9.427/1996, 10.847/2004 and 10.848/2004, new projects have been granted, structuring the sector into state-controlled and verticalized operating companies, operating in the generation, transmission and distribution of energy, facilitating access to energy for the population (BRASIL, 1996; BRASIL, 2004a and BRASIL, 2004b).

In 2015, electricity consumption in the country was 522.8 TWh (EPE, 2016), with a forecast of 693.47 TWh in 2024 (EPE, 2015b). Regionally, the highest residential consumption was in the Southeast, naturally because it is the country's most populous region (FERREIRA, 2016) (Figure 1).

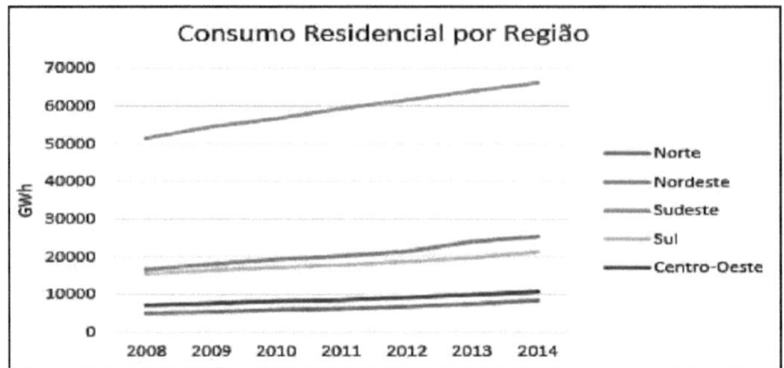

Figure 1. Residential electricity consumption by region. Source: EPE, 2016.

In 2015, Brazil produced 581 TWh of electricity, which is 1.52% less than the previous year (EPE, 2016). In 2015, public service plants were responsible for generating 84.1% of this energy and the remaining 15.9% was generated by self-producers (EPE, 2015b).

According to the world energy balance, the country's main source of electricity is hydroelectric, although it fell by 1.84% compared to 2014, it still accounts for 64% of the Brazilian energy matrix (EPE, 2016) (Figure 2).

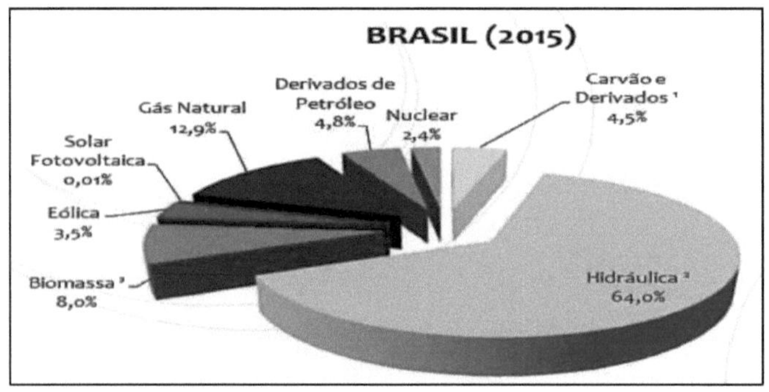

Figure 2. Brazil's main electricity generators in 2015. Source: EPE, 2016.

Brazil's installed electricity generation capacity grew by 7,936 MW in 2015, to a total of 140,858 MW, with the determining factors for this expansion being the installation of hydroelectric power stations, thermal power stations and wind and solar power plants, making it possible to increase generation capacity (EPE, 2016) (Table 2).

Table 2. Electricity generation capacity in Brazil.

Source	2015	2014	15/14
Hydroelectric	91.650	89.193	2,8%
Thermal	39.564	37.827	4,6%
Nuclear	1.990	1.990	0,0%
Wind	7.633	4.888	56,2%
Solar	21	15	42,3%
Available capacity	**140.858**	**133.914**	**5,2%**

Source: EPE, 2016.

Given this generation capacity, the main energy-consuming sectors in 2015 were the industrial, transportation, energy and residential sectors (EPE, 2016) (Figure 3).

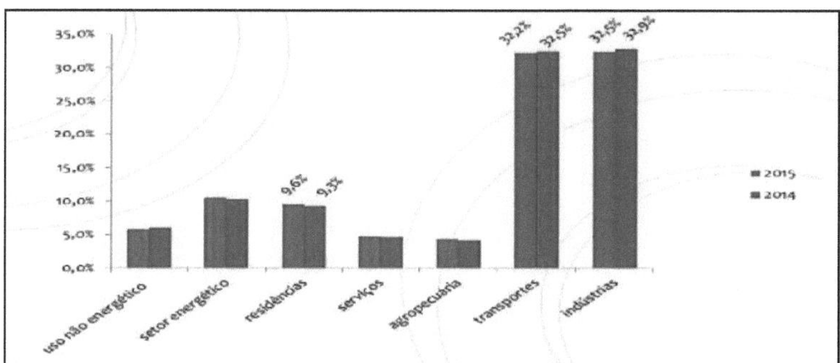

Figure 3: Electricity consumption in Brazil between 2014 and 2015. Source: EPE, 2016.

2.2. TRANSMISSION LINES

With the discovery of electricity, the need arose to transmit the energy generated to the distribution systems, because in Brazil, most of the hydroelectric plants are far from the large consumption centers, making it essential to make an extensive and reliable transmission network viable, capable of transporting all the available supply, uniting resources and optimizing energy generation in the country (MENEZES, 2015).

In this way, transmission lines (TLs) are responsible for transporting energy to various regions of the country, connecting not only power plants to large consumers, but also those who purchase energy at high voltages, such as factories and mining companies, or energy distribution companies, which are responsible for transporting the energy to smaller consumers (ABRADEE, 2016).

The TL is an electrical circuit that interconnects different types of substations, consisting of metallic conductor wires suspended from towers (self-supporting or guyed), by means of ceramic insulators or other highly insulating materials (ABRADEE, 2016).

The alternating current (AC) system uses three-phase networks with one or more subconductors per phase and is the most widely used because it is more flexible, allows electricity to be generated, transmitted, distributed and used at the most economical and safe voltage, while direct current (DC) transmission is used to transport large amounts of power over long distances because it has lower costs and losses (MENEZES, 2015).

In order to be called a "Transmission Line", the transportation of electricity, i.e. the voltage, must be higher than 138kV, because below this value we have sub-transmission and distribution lines (SANTOS, 2011). Transmission, in turn, can be carried out through overhead, underground or underwater lines, where conventional conduction is carried out through overhead lines (MENEZES, 2015).

Brazil has a system for generating and transmitting energy made up of power plants, transmission lines and distribution assets, the so-called National Interconnected System (SIN), known as the "electric highway", which covers a large part of Brazilian territory, where the installation of this network has evolved according to the demand of the Brazilian regions (SANTOS, 2011).

Currently, there are programs to expand the SIN, with the intention of setting up new power generation and transmission facilities to meet the growth of the consumer market. This incentive is managed by ANEEL, through auctions to select groups of entrepreneurs responsible for building and operating these networks (ANEEL, 2005).

Planning for the expansion of the country's transmission system is carried out jointly with the Energy Research Company (EPE) and the National Electricity System Operator (ONS). The projects defined by the Federal Government are included in the National Privatization Program (PND), which ANEEL determines to promote and monitor the bidding processes for the respective concessions (ANNEL, 2005).

Brazil's complex transmission network already exceeds 116,000 km (EPE, 2014) and is considered the largest interconnected network in the world (ELETROBRAS, 2015). This system electrically connects all regions of the country, except for small isolated systems in the Amazon or systems of a private nature, which account for a small 1.7% of the total installed (ONS, 2015).

However, even though most of the country is served by transmission networks, the growing demand requires the permanent expansion and reinforcement of the network, so that there is quality and reliability in the service (MENEZES, 2015).

Given the importance of transmission line projects, estimates show that by 2019 the electricity sector will receive an investment of approximately 20 billion reais, 30% of which was in 2012 (EPE, 2015b).

However, the construction of transmission lines, whether for national interconnection or just to meet specific demands, is a service that requires many technical, economic and environmental feasibility studies at all stages of the project (MENEZES, 2015).

2.3. ENVIRONMENTAL IMPACT AND ENVIRONMENTAL LICENSE

The industrial revolution and population growth have led to changes in society, intensifying changes to the environment through the exploitation of natural resources (ALMEIDA et al., 2010). As a result, the demand for energy resources increased, leading to the disorderly consumption of fossil fuels and other natural energy sources, as humanity realized that the quality of life was improving as a result of energy production and consumption (TRIGUEIRO, 2005).

However, various environmental impacts arise from energy transmission. It is therefore essential to comply with CONAMA Resolution 237/97, which defines activities and/or undertakings that require specific licensing for the construction, installation, expansion and operation of establishments and activities that use environmental resources, which may cause environmental impacts of varying degrees of complexity (BRASIL, 1997).

In the case of electrical projects, licensing is supervised jointly by municipal, state and federal agencies through IBAMA. Normally, the licensing of large projects involving impacts on more than one state, such as transmission lines, oil and gas activities on the continental shelf, are licensed by IBAMA.

Resolution 001/86 establishes definitions, responsibilities, criteria and general guidelines for the use and implementation of Environmental Impact Assessment. Article 2 of this resolution clarifies that electricity transmission lines with a capacity above 230 Kv will depend on the preparation of an environmental impact study and respective environmental impact report - RIMA, to be submitted for approval to the competent state body and IBAMA (BRASIL, 1986).

Law 6.938/81 and CONAMA Resolutions 001/86 and 237/97 are the basic guidelines for compliance with environmental licensing (BRASIL, 1981; BRASIL, 1986; BRASIL, 1997). In addition to these, the Ministry of the Environment recently issued Opinion No. 312, which discusses state and federal competence for licensing, based on the scope of the impact (BRASIL, 2004c).

Decree 97.632/89 defines degradation as processes resulting from damage to the environment, whereby some of its properties are lost or reduced (BRASIL, 1989). Thus, in order to comply with Law 6.938/81, it is mandatory to draw up a Degraded Areas Recovery Plan (PRAD) for electricity transmission projects, with the aim of preserving, improving and recovering environmental quality (BRASIL, 1981).

Article 23 of Law 8.171/91 clarifies that companies that exploit electricity will be responsible for the environmental alterations caused and thus obliged to recover the environment in the area covered by the project (BRASIL, 1991), since it is a crime under Law 9.608/98 not to recover the exploited area (BRASIL, 1998). Law No. 18.104, of July 18, 2013, establishes the new Forestry Policy for the State of Goiás and makes other provisions, including general rules aimed at protecting and improving environmental quality in the territory of the State of Goiás (BRASIL, 2013).

In view of this, the PRAD must gather information, diagnoses, surveys and studies that allow the assessment of the degradation or alteration and the consequent definition of appropriate measures to recover the area (BRASIL, 1981). Normative Instruction No. 4, drawn up by the Brazilian Institute for the Environment and Renewable Natural Resources, establishes all the procedures for drawing up this program (BRASIL, 2011a). Allied to PRAD, where appropriate, the methodology for recovering Permanent Preservation Areas (APPs) is used, taking into account the criteria set out in CONAMA Resolution No. 429/2011 (BRASIL, 2011b).

Therefore, once the changes to the environment have been diagnosed, it is essential to reverse or compensate for the impacts. Therefore, all the recovery techniques presented in this report are in accordance with IBAMA Normative Instruction No. 4 of April 13, 2011 (BRASIL, 2011a), CONAMA Resolution No. 429 of February 28, 2011 (BRASIL, 2011 b), as well as the Basic Environmental Plan, P.03, prepared by Belo Monte Transmissora de

Energia SPE S.A. in conjunction with JGP Consultoria e Participações Ltda., which establishes all the items that the program must contain (PBA, 2015).

15

3. ACTIVITIES CARRIED OUT

3.1. GENERAL INFORMATION

The Compulsory Supervised Internship was carried out at Sao Simao Montagens e Serviços LTDA, part of the Cobra Group, located at GO 330, Km 153, sector ring road, in the municipality of Ipameri-GO. It took place from July 20, 2016 to November 21, 2016, in the area of Electricity Transmission, under the guidance of Forestry Engineer Gustavo Pereira Nunes, totaling 528 hours.

3.2. COMPANY DESCRIPTION

Sao Simao Montagens e Serviços, headquartered in Rio de Janeiro and located in Ipameri-GO, is a Spanish power transmission company belonging to the Cobra Group, present in 45 countries. It carries out various activities, but its main activity is the construction of electricity distribution stations and networks.

The company's mission is to offer services to small and large clients, owners or concessionaires around the world, in order to create and operate industrial infrastructures, processes and technologies.

Since its creation in 2007, it has carried out activities related to energy transmission, canal construction, commercial lines, pipelines and dredging. In its 9 years of existence, it has 3,673 establishments and a large and complex corporate structure, with the development of numerous projects of socio-economic relevance to various regions of the country (GRUPO COBRA, 2016).

The company is currently one of those responsible for the LT ± 800 kilovolt (kV) CC Xingu/Estreito transmission line corresponding to Lot AB of ANEEL's Auction 011/2013, awarded to the company by Belo Monte, the company primarily responsible for the work (ANNEL, 2013). The transmission line is the first ultra-high-voltage line with unprecedented technology in Brazil, as it allows the transportation of energy with reduced losses, and is planned to take the energy generated by the Belo Monte Hydroelectric Power Plant (HPP) to the Southeast Region, with more than 2,086.9 km of extension, which should be completed by 2018 and foresees investments of R$ 5 billion (BELO MONTE, 2016).

In addition to the transmission line, the project also includes associated facilities, including two converter stations (from direct current to alternating current and vice versa), two earth electrodes installed in the municipalities of Anapu (PA) and Altinópolis (SP) and two electrode lines that connect the electrodes to the converter stations. One of the electrode lines will be installed in the municipality of Anapu and the other will intercept the territories of Ibiraci and Claraval (MG) and Franca, Patrocinio Paulista and Altinópolis (SP) (BELO MONTE, 2016).

The TL intercepts 65 municipalities in four states: Parà, Tocantins, Goiàs and Minas Gerais. In Goiàs, the company intercepts 23 municipalities Porangatu, Santa Tereza de Goiàs, Estrela do Norte, Mara Rosa, Campinorte, Nova Iguaçu de Goiàs, Uruaçu, Santa Rita do Novo Destino, Barro Alto, Vila Propicio, Cocalzinho de Goiàs, Corumbà de Goiàs, Alexânia, Abadiânia, Silvânia, Vianópolis, Orizona, Urutai, Ipameri, Campo Alegre de Goiàs, Catalao, Ouvidor and Très Ranchos.

Belo Monte divided the TL into 8 sections and awarded construction responsibility to the four companies, with Sao Simao being responsible for two of them, section 6 and 7, which are about 500 kilometers long. Section 6 covers the municipalities of Uruaçu and Cocalzinho, while section 7 intersects the municipalities of Ipameri - main construction site, Silvânia, Orizona, Pires Belo and Alexânia - auxiliary construction sites.

In the process of building the TL, various environmentally damaging activities are observed, such as the installation of accesses, clearing the right-of-way, foundations, erection of structures, grounding and laying cables. In view of this, there is a notable need for forestry engineers who are experts in monitoring these impacting activities.

3.3. INTERNSHIP PRACTICES

Various activities were carried out during the internship, such as drawing up cubing reports, monitoring the fumes released by diesel vehicles, meeting the requirements of inspectors, carrying out environmental training and building a temporary nursery. However, the activity that took the most time was the preparation of a Degraded Areas Recovery Program, the emphasis of which will be given in this report.

3.3.1. Geoprocessing

Geoprocessing activities were monitored, suppression maps drawn up, access sketches defined, points allocated in the field and on Google Earth in order to define the best access routes to the towers and plan the execution of suppression-related activities. All the suppression routes were carried out in a specific program and defined together with the landowners, because as the line would intercept their properties, they had to advise us on the best routes. Once these sketches had been drawn up, they were presented to the landowners so that they could give a favorable opinion. Figure 4 below illustrates a sketch defining access routes.

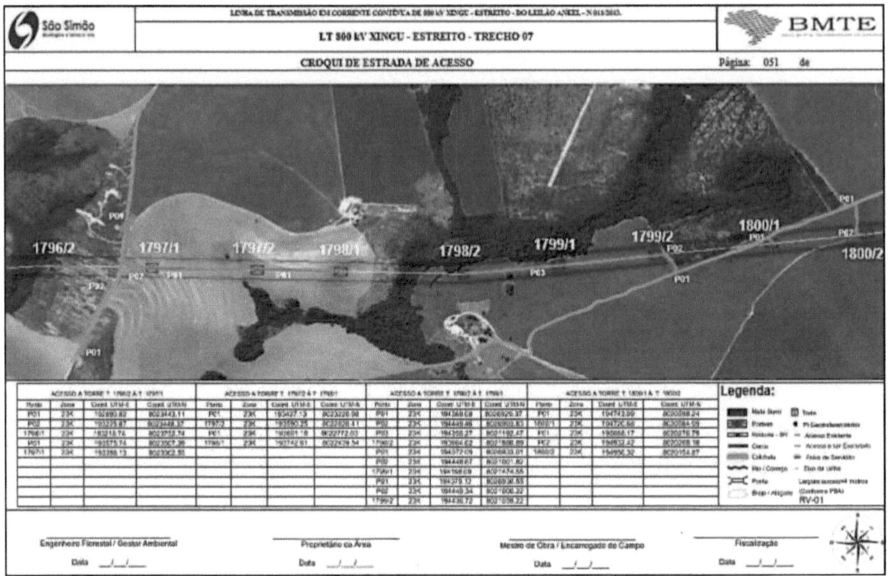

Figure 4 - Sketch defining access routes. Source: Personal archive, 2016.

3.3.2. Diesel smoke control

The construction companies providing services for Belo Monte, such as Sâo Simao, are obliged to control the emission of pollutants during all stages of the work, in order to reduce the negative impact on air quality in areas with nearby residences, provide comfort for the workers, collaborate in maintaining air quality and prevent accidents on the construction site.

Therefore, smoke monitoring was carried out on around 157 vehicles, corresponding to the company's entire fleet, using the Ringelmann scale according to ABNT through NBR's 6016, NBR 6065 and NBR 7027 (ABNT, 2015); (ABNT, 2001) and (ABNT, 1980) (Figure 5). This scale has 5 levels of smoke coloration, where all vehicles with smoke levels 3, 4 and 5 were sent for maintenance in order to minimize pollutant emissions.

Figure 5 Smoke monitoring. A) Quantification; B) Scale of levels. Source: Personal archive, 2016.

3.3.3. Environmental monitoring

The company has ISO 14001 environmental management, ISO 9001 quality and OSHAS 18001 occupational health and safety seals, and is constantly being audited by inspection companies such as Marmoré and JGP, which is why it is necessary to draw up monthly Proof of Compliance documents.

Therefore, during the internship it was necessary to study various documents, such as the Basic Environmental Plan (PBA), the Vegetation Suppression Authorization (ASV), the Environmental Impact Study (EIA) and other legal documents, in order to draw up monthly reports on the company's environmental performance.

The documentation included Black Smoke Monitoring, the Forest Origin Document - DOF, the License to Use and Carry Chainsaws - LPU, environmental licenses (Previous License - LP; Installation License - LI and Operating License - LO) from the company's raw material suppliers, among others.

3.3.4. Plant suppression

The construction of a power transmission line requires the execution of various activities, which are usually related to deforestation, such as the implementation of accesses, clearing the right-of-way, founding the towers, assembling the structures, grounding and laying cables.

The opening of accesses represents all the services related to the opening of roads and access paths, responsible for connecting the areas where the structures are located to the nearest roads, and it is essential that vegetation is cleared. Deforestation is also necessary in the tower areas, in the span and in the right-of-way, in order to guarantee the

19

safety of the TL and the people who live near it.

In this way, plant suppression is inevitable if the process is to continue, because only after deforestation is it possible to lay the foundations and erect the towers, which are responsible for supporting the electrical circuit and the stresses caused by the cables and the action of the winds. During the grounding and cable laying phase, there is no actual suppression, but rather the selective cutting of certain species that could damage the power transmission network in the short and long term.

In view of this, deforestation is present at all stages of the construction of power transmission lines. Thus, during the internship, a suppression control spreadsheet was developed, capable of controlling, over time, the towers that went through suppression, the species in extinction and the volume of wood removed, in order to proportionally replace the removals in the future.

Visits were also made to the field to monitor plant suppression, the removal of native timber and commercial planting, which made it possible to get to know a large part of the activities carried out by the company in the field. Below is a record of one of the areas of vegetation suppression monitored during the internship (Figure 6).

Figure 6 - Plant suppression. A) Access opening; B) Tower area opening. Source: Personal archive, 2016.

3.3.5. Forest inventory

After the deforestation and removal of the woody material, cubing reports were drawn up, a kind of inventory form for each tower and span. This document contained some information on species characterization, the amount of wood removed and details of the owners and the property where the removal took place. Figure 7 shows an area with timber girdling that was prepared to be inventoried

These documents are essential for controlling and reporting on plant suppression to inspection companies and IBAMA. They are also required for the donation of woody material to interested owners.

Figure 7. Harvesting wood for a forest inventory. Source: Personal archive, 2016.

3.3.6. Nursery construction and seedling production

The extensive deforestation and consequent environmental impacts caused by the transmission lines require mitigation measures. One of the relevant activities proposed and carried out during the internship was the setting up of a temporary nursery, with the aim of producing native seedlings for the recovery of these impacted and adjacent areas.

The nursery project defined all its dimensions, the materials needed and its seedling production capacity, which was around 3,600 seedlings.

The seeds acquired for the production of seedlings were collected from areas adjacent to those impacted, with the aim of returning the region's species in the form of seedlings. In addition, each species was induced to break dormancy according to its physiological particularity, providing uniform and quality development. Figure 8 shows the general appearance of the nursery and the native seeds after processing.

Figure 8. Forest nursery. A) General view; B) Seed processing; C) Native seedlings. Source: Personal archive, 2016.

3.3.7. Program to recover degraded areas

The aim of the program is to comply with legal requirements regarding the recovery of the area that suffered environmental impact as a result of the installation of tower number 1757/1 in the municipality of Ipameri-GO (Figure 9). Environmentally correct techniques will be used in this area to enable environmental balance and compensate for the damage caused by the occupation of the enterprise.

Figure 9. A) General view of the tower site in Ipameri-GO; B) Environmental impacts. Source: Personal archive, 2016.

The main activities carried out during the internship consisted of applying the degraded area recovery program (PRAD) to the area of the project mentioned above, as detailed below.

3.3.7.1. Description of the project

3.3.7.1.1. General Information

Name or corporate name of company: Sâo Simao Montagens e Serviços Ltda; CNPJ: 08.885.461/0001-09; Address of company: GO 330, Km 12, Anel Viàrio, Ipameri-Goiàs; Type of activity: Construction of electricity distribution stations and networks.

3.3.7.2. Basic information about the project

The project consists of a power transmission line with 800 kV power, where it is necessary to carry out various activities, such as setting up accesses, removing and clearing the right-of-way, founding the towers, assembling the structures, earthing and laying the cables.

This process of installing the project generates various environmental impacts in each of the activities carried out, which causes alterations to the environmental balance. However, from a technical point of view, most of the changes caused are considered to be recoverable.

3.3.7.2.1. Implementing access

They represent all the services related to the opening of roads and access routes, responsible for linking the areas where the structures are located to the nearest roads. They are built with a width of four meters, sufficient to allow the passage of vehicles and large equipment needed to carry out the activities. These roads can be considered temporary or permanent.

3.3.7.2.2. Felling and clearing of the right-of-way

For the power transmission line to pass through, it is necessary to remove vegetation in the areas around the towers, the shaft and the right-of-way, in order to guarantee the safety of the transmission line and the people living near it. This area is defined according to the swaying of the conductor cables due to the action of the wind, the electric field, radio interference, noise and the position of the foundations (BELO MONTE, 2016).

The easement strip is defined by the Brazilian Association of Technical Standards ABNT through NBR 5.422/85 (ABNT, 1985). In this project, the easement strip has a width of 110 meters, 55 on each side of the route axis, promoting satisfactory conditions for construction, operation and maintenance of the line (MENEZES, 2015).

3.3.7.2.3. Tower foundations

The foundations of the towers are built to support and transfer the loads of the structures to the ground. In the case of transmission lines, this stage is fundamental, as the towers need to be fixed directly to the ground in order to withstand the stresses caused by solicitous efforts (MENEZES, 2015). Before the foundations are laid, earthworks are carried out to shape and level the ground in order to facilitate the other phases of construction.

3.3.7.2.4. Assembling the structures

Structures are basic components of the transmission network made of steel lattice material that have the function of physically supporting the electrical circuit, in addition to withstanding the stresses caused by the cables by the action of the winds, maintaining an ideal spacing between conductor cables and lightning rods. Together with the cables and foundations, the structures are the most labor-intensive part of overhead transmission line construction (MENEZES, 2015).

The assembly of towers in the field is divided into three stages: pre-assembly, assembly and overhaul of the tower. Pre-assembly is nothing more than spreading out and positioning the tower parts on the ground and is intended to make it easier to hoist them up at the time of actual assembly, which is carried out with the help of a wheeled tractor. After assembly, however, the overhaul is carried out, where the final tightening is done with the help of a torque wrench and checks that all the parts are correctly installed.

A total of 3,749 towers have been allocated to this project, 80% guyed and 20% self-supporting. The clearing area for each tower is different, with self-supporting towers having an average area of 2,500 m^2 (50 m x 50 m), while guyed towers have an average area of 3,000 m^2 (60 m x 50 m) (BELO MONTE, 2016).

3.3.7.2.5. Grounding

The grounding system is implemented by the company because it promotes the discharge of any induced electric current to earth, preventing it from spreading. It is made up of counterweight cables and impedances, installed both at the foot of the anchor towers and at the base of the central mast and pylons of guyed structures.

Therefore, its purpose is to keep people and animals that may touch it safe, since fences that cross or pass through TLs can be energized by the effect of electromagnetic induction (MENEZES, 2015).

3.3.7.2.6. Cable laying

After assembling the structures and grounding, the installation stage of the lightning cables and conductors begins, which includes laying, splicing, bending and stapling in accordance with the technical specifications and safety standards. The cables are laid on the basis of the Laying Plan, which assesses all the conditions and obstacles along the route of the TL in order to find the best distribution of the coils in the field (MENEZES, 2015).

In the extra-high-voltage transmission line, the cables are laid under controlled tension, i.e. a lighter steel cable (pilot cable) is first laid and then connected to the pilot by means of a rocker arm. The cables are then pulled by a winch located at the end of the section called the winch square, while at the other end (the brake square) the cables leave the reels and pass through the brake, where the tension of the launch is controlled.

3.3.7.3. Location and characterization of the degraded area

The route of the Xingu/Estreito Transmission Line in Ipamero territory is 60 km long, and the area directly affected covers around 0.25 hectares and 8.72 m^3 of native wood.

The area that will be subject to the recovery plan is characterized by the installation of a power transmission tower on the property of Mr. Nissauro Pereira Duarte, in the municipality of Ipameri, whose geographic coordinates are 17°31'1.52"S, 48°0'36.33"O, and at 827 m altitude, it is located 1.39 km from the owner's house, and 33.1 km from the nearest neighborhood (Figure 10).

Figure 10. Location of the development. Source: Google Earth, 2016.

During the process of installing the TL on this tower, environmental liabilities were assumed. In this context, the main activities responsible for causing environmental degradation were the removal of vegetation from the hill in order to build accesses, clearing the right of way, digging for the foundations of the towers, assembling the structures, earthing up and laying cables. Leaving the area exposed to various environmental impacts.

3.3.7.4. Environmental impacts

The construction of TLs provides a number of socio-economic benefits, bringing greater comfort to society. However, despite the economic importance of this type of project, the environmental impact is inevitable.

Before starting work, IBAMA carried out a study on the most appropriate route in terms of the environment, through territorial evaluations, taking into account Conservation Units, Priority Areas for Biodiversity Conservation, Settlement Projects, Indigenous Lands, and caves. It also took into account interference with watercourses, floodplain crossings, areas with unfavorable geomorphology, fragments of medium-sized native vegetation, urban areas and urban expansion (PBA, 2015).

In view of this, the route was laid out in such a way as to minimize environmental impacts, so wherever possible it was placed in areas that had already been deforested or degraded. Even with this strategy, the impacts are inevitable.

It is undeniable that new developments cause environmental changes, which can be positive or negative. Therefore, in order to minimize the negative effects and enhance the positive ones, an environmental impact analysis was carried out, seeking to identify and evaluate all possible impacting activities, facilitating decision-making to minimize them.

In order to identify the impacts and their origins, studies of the physical, biological and socio-economic environments were carried out. Table 3 details the main impact-causing activities carried out by the company throughout the process of installing the TL, as well as the tools or machinery used in the respective activities.

Table 3. Description of the activities involved in setting up the Transmission Line, which may generate environmental impact, but which can be recovered and mitigated.

Activity	Description	Tools or machinery used
Opening Access Routes	Construction or renovation of roads linking the site of the TL to the main roads in the region	Backhoe loaders.
Clearing the right-of-way	Suppression of tree species present in forest fragments intercepted by the TL's right of way	Chainsaws and tractors with trailers
Excavations for the	Opening the foundations, where the tower	Backhoe loaders, shovels, drills, hydraulic auger trucks,

foundations	support bases will be fixed	hand diggers and picks
Tower assembly	Assembly of parts up to full installation of the tower structure	Munck truck, wrench, tractor with winch and crane
Launch	Power generation cable launches	Tractor with winch, wheel loader and munck truck

The environmental effects observed during the implementation phase were: impacts on flora, fauna, soil and water resources, as described below:

3.3.7.4.1. Impacts on flora

The impacts on the flora are mainly due to the need to remove native vegetation in order to install the accesses, open the right-of-way and consequently provide the conditions to carry out other activities, such as foundations, assembly and launching, since the area that the line intercepts must be devoid of vegetation in order to guarantee the safety of the line after it is energized (FERREIRA, 2011).

During the operation of the line, it is essential to mow the regenerating native vegetation in the areas of the towers, tracks and accesses that will continue to be used, in order to keep it at a height that does not put the structures and conductor cables at risk. In addition, whenever necessary, pruning and selective cutting of trees that pose a risk outside these areas should be carried out.

Deforestation took place within the limits proposed by Plant Suppression Authorization No. 1,082/2015, in an attempt to minimize the impacts caused. Even so, the suppression causes loss and fragmentation of germplasm, which reduces the availability of habitat for local fauna and affects the balance of flora. In general, several species are affected, especially in terms of the balance of these habitats.

In addition, suppression implies the emergence of new edges in the remaining formations, exposing portions that were previously protected inside. Depending on the characteristics of the vegetation, the creation of new edges can cause changes in structure and composition (known as the "edge effect").

Another impact is the increased risk of fires in the remaining adjacent vegetation. This risk is related to the increased traffic of machinery, equipment, workers and accidental electrical discharges in areas of native vegetation.

3.3.7.4.2. Impacts on fauna

Interference with fauna as well as flora is essentially associated with deforestation, since the removal of vegetation alters the characteristics, reduces the habitat and directly influences the local fauna, through a reduction in the number of animals and structural changes in the faunal communities caused by the edge effect (FERREIRA, 2012).

Depending on the faunal group, plant suppression and habitat alteration can act with greater or lesser intensity, since larger forest fragments can be a matrix for wild populations and tend to have greater faunal biodiversity than smaller ones.

The removal of woody material results in an increase in the incidence of light, wind, temperature variations and a decrease in humidity, and so these items end up interfering with animal behavior, a fact that can be perceived up to 500 meters from the edge of the forest, also known as hidden effects (CAMPOS, 2010).

The local noise caused by the movement of machinery and the movement of workers could induce the displacement of various species of fauna and other vertebrates present in the vicinity of the works to adjacent regions.

The high number of workers on the construction sites can contribute to increased hunting pressure, especially in the more preserved areas, as well as the opening of service roads, including by the surrounding population, due to easier access.

Accidents with terrestrial fauna during construction can occur, especially during excavation activities for the tower foundations and vegetation clearance, and also due to the increase in construction-related traffic. The presence of synanthropic and opportunistic animals such as dogs, cats, pigeons and rats can increase in construction site areas. This can result in an increased risk of disease transmission to wildlife (epizootics), thus increasing the chances of disease transmission between wild species.

Another impact is the "conflict" between the animals and the electricity cables, which occurs when the animal comes into contact with an energized part of the network, receiving an electrical discharge which causes death. In addition, there is the collision and fall of birds on the wiring, which has not yet been recorded, as the line has not yet been completed.

3.3.7.4.3. Impacts on soil

The main impacts on the soil come from the removal of vegetation, where the fertile layer is eliminated and the soil remains bare throughout the construction period. On the access roads, the situation is a little more critical, as they will remain bare. In addition, the constant passage of heavy machinery aggravates the physical conditions of the soil and all these factors contribute to soil erosion and compaction.

Other interventions have been to alter the terrain, making it unstable, as it already has a very steep slope and deforestation has induced erosion processes. In addition, there have been oil leaks from the machines, which, although small, cause contamination of the soil and even water resources.

3.3.7.4.4. Impacts on water resources

All areas that have undergone deforestation are subject to changes in the water infiltration and drainage system. In the area under study, the situation is more delicate, as the slope facilitates soil runoff and consequently, in the long term, can silt up water resources.

Siltation is responsible for an increase in the turbidity of surface water, since without soil cover, mineral particles are easily carried away by rain, which leads to an increase in solid material above the river's transport capacity. This creates new flows and velocities, as well as changing the dynamics of surface runoff, which must be controlled (CAMPOS, 2010).

3.3.7.4.5. Impacts on the air

The main impacts on the air were the suspension of dust due to land clearing, the movement of earth from the foundations, earthworks, the implementation of accesses and the circulation of vehicles on the unpaved roads.

3.3.7.4.6. Magnitude of environmental impacts

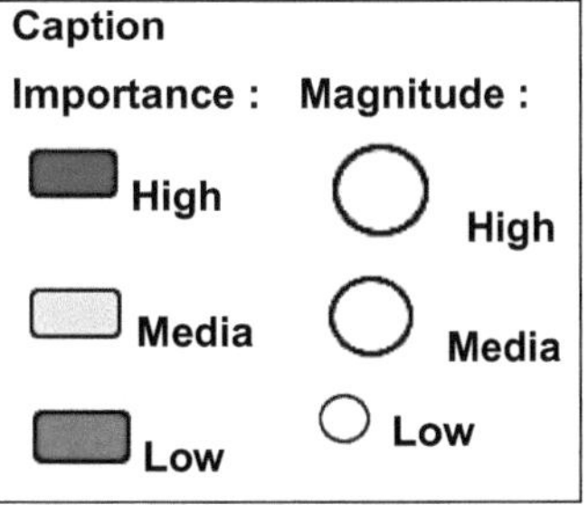

<u>**Table 4** - Magnitude of environmental impacts observed with the PRAD.</u>

Environmental impacts	Importance and Magnitude	
	Implementation	Operation
1. Impacts on Flora		
1. Reduction of vegetation cover and/or habitats	●	
2. Reduction in the population of individual species protected and/or endangered	●	
3. fragmentation and/or altered connectivity between remnants of adjacent native vegetation	●	●
4. alteration of the remaining adjacent vegetation due to the edge effect	○	○
5. Increasing the risk of fires in the area adjacent remnant vegetation	○	○
6. Risk of inducing the exploitation of vegetation adjacent remnant with the creation or improvement of land accesses		●
7. inhibition of secondary vegetation regeneration in the right-of-way during the operation of the LT		●
2. Impacts on wildlife		
1. reduction of living areas for fauna species local	●	
2. scaring away wildlife during construction	○	
3. increased risk of hunting	●	
4 Accidents involving terrestrial fauna during the construction	●	
5. bird accidents during operations		○

6. attracting synanthropic fauna during construction	●
3. Impacts on Soil	
1. landform alteration, slope instability and inducing erosive processes	○
2. risk of soil contamination	○
4. Impacts on Water Resources	
1. increased turbidity in watercourses	○
2. siltation of watercourses	○
3. risk of contamination of drains	○
5. Impacts on the Air	
1. Changes in air quality on construction sites during construction	○

Source: Personal archive, 2016.

3.3.7.5. Mitigation measures

The legal obligation to recover and preserve the natural environment affected by the construction of the project justifies the adoption of PRAD. Through this program, the effects of the implementation of the TLs on flora, fauna, water resources, soil and air can be mitigated or even eliminated.

Methodologies for preventing and mitigating environmental impacts have been drawn up and proposed for all activities associated with the transmission lines, seeking out the best techniques. In addition, these measures will be carried out after the company's activities have ended, in accordance with the environmental actions for transmission systems drawn up by JGP, which apply to São Simão Montagens e Serviços.

3.3.7.5.1. Soil

Soil plays a direct role in the development of plants, microorganisms and water reserves, as well as being one of the vital components of the environment (FERREIRA, 2011).

The opening up of access roads to the structures is one of the biggest environmental impact factors in TL projects, since suppression and consequent interference with the soil are inevitable (FERREIRA, 2011). However, the removal of vegetation throughout the project is minimal, as existing accesses are used whenever possible.

Even so, the 1757/1 tower suffered impacts related to the opening of the access, due to the steep slope it is on, combined with the texture of the soil, making it a victim of erosion and accelerated wear and tear of the soil, compromising its physical, chemical and biological composition.

In view of this, conservation practices will be adopted in order to prevent erosion processes, as well as contributing to the restoration of the Cerrado area, which has been de-characterized by the direct action of the project. These operational actions seek to restore balance to the destabilized area, reduce soil loss and prevent silting of the drainage network.

Initially, the site will be cleared of debris and other remains of construction materials that may have weathered over time, such as gravel, leftover concrete, plastic bags, hardware and so on. This ensures the best conditions for re-establishing the vegetation.

The materials will be removed manually with shovels and hoes, as they are found in small quantities. They will then be sent to an appropriate location that has been previously established.

Subsequently, the area will be cordoned off with fences and information signs, so that there is no interference of any kind, thus preventing animals or people from passing through and damaging the germination of the seeds and the development of the plants, as well as compacting the soil.

Terraces will be built, with varying spacing, taking into account the slope and texture of the soil. This technique prevents erosion, improves the visual aspect of the terrain and provides stability. Ditches and drainage channels will also be built to allow the normal flow and dispersal of rainwater, keeping them clean and unobstructed at all times.

However, in order to make the implementation of storm drainage and revegetation plans feasible, it is necessary to smooth out the terrain using machinery, which provides a more uniform slope, thus reducing the speed of the water and, consequently, the production of sediment.

After all of these stages, we will use the transposition of the branches in the form of beds, which provides use of the waste, energy dissipation and protection against the impact of raindrops, while at the same time the beds can germinate or regrow and provide organic matter to the soil.

At the same time, this transposition provides a microclimate that is favorable to the proliferation of some insects such as wood-degrading coleopterans (termites) and attracts various animals such as rodents, snakes and birds. In this way, this technique has the effect of rescuing flora and fauna (REIS et al., 2005).

The planting of native vegetation will be carried out without interfering with the safety of the line. It is also a simple conservation practice that will ensure the preservation of the soil and allow the local water table to be replenished, which is why it will be used in this program.

With regard to soil contamination by combustible waste, environmental training is carried out and environmental kits are made available to collect this contaminating material.

3.3.7.5.2. Flora

After using soil conservation techniques, green manuring will be carried out using leguminosae, with the aim of recovering the quality of the soil, flora and fauna.

This will be followed by the re-establishment of vegetation, a fundamental measure that allows for greater infiltration, less surface runoff and also protects the soil against laminar erosion (PEREIRA et al., 2008).

During the internship, seeds were collected from the adjacent vegetation and sent to the forest nursery to produce seedlings that could be used for revegetation.

Revegetation uses herbaceous species from the region. Thus, the planting will consist of a simultaneous mixture of different species of native and pioneer grasses and legumes, with root systems capable of interlacing for better erosion control.

The planting will be in islands of high diversity, made up of plants with different life forms and adaptations to successional stages, pollination and dispersal processes. Distributed throughout the year, the technique has been used in environmental programs to facilitate the recovery of degraded areas.

Revegetation will essentially take place in the tower squares, launches, accesses and other relevant locations, respecting the site's safety constraints. As the area in question is sloping, simultaneous planting of seedlings, stolons or slabs will be used. Stoloniferous species are suitable for areas subject to erosion, such as hillsides, as they quickly cover the soil, trapping particles and protecting against the direct impact of rain and sun. In addition, they are able to close the gaps between plants, preventing weeds from invading (CORRÊA, 1996).

As for fertilization, the transposition of branches together with green manure will be able to have important effects on the development of the seedlings. At the same time, the use of herbaceous species with deep root systems can help to decompress the soil by breaking up compacted layers.

3.3.7.5.3. Fauna

In degraded areas, the absence of vegetation is associated with a lack of seeds due to the impacts of plant suppression. Without it, there are no more environments for shelter and food for the animals capable of dispersing the seeds.

In view of this, direct methods capable of attracting local fauna will be adopted. It is important to note that the transposition of branches and reforestation indirectly promote conditions for the survival of these animals.

Perches will be set up to allow animals from nearby fragments to sporadically come into the degraded areas and consequently bring back seeds that will be dispersed in the area (FERREIRA, 2011).

The dry perches imitate dry tree branches for the birds to perch on and will be made from various materials such as wood scraps or bamboo and a fast-growing species will be grown at their base. In addition to the perches, aerial cables will be installed linking the perches. These cables will be made from rope or any similar material available.

Another form of attraction is the planting of fruit trees, which are attracted by these living perches that provide food (REIS et al., 2005). Over time, the plant develops and the perch becomes a protected environment, providing shelter for the animals.

Whenever appropriate, the animals will be rescued and relocated to a nearby habitat with similar characteristics to the environment they were in.

3.3.7.5.4. Air

As for changes in air quality at construction sites during construction, the suspension of particles is controlled by humidifying the roads with water trucks, and another measure is to make drivers aware of the need to reduce their speed.

3.3.7.5.5. **Water resources**

The risks of silting up watercourses will be contained simultaneously through the reforestation and soil conservation techniques mentioned above.

As for the risks of contamination, various forms of employee awareness-raising are carried out through environmental training and dialog to avoid these impacts.

4. FINAL CONSIDERATIONS

The supervised internship provided a unique learning opportunity, as it made it possible to carry out practical activities involving a wide range of subjects related to environmental issues. By accompanying them in the field, supervising them and drawing up environmental reports, it was possible to integrate theoretical content learned during the degree course, which is fundamental to professional training.

The work carried out provided privileged academic experience, through knowledge of various forms of professional activity in the field of forestry engineering, experience of working in large groups, as well as routine situations in the context of an electricity transmission line company.

In addition, by characterizing all the environmental impacts resulting from the installation of the LT Xingu-Estreito power transmission line, it was possible to present the proposed solutions in a clear and viable manner. Studies of this nature, in the short and medium term, correct the problems arising from inappropriate land use and, in the long term, restore part of the ecological processes that were once lost, restructuring plant and animal populations and making the areas to be recovered autonomous and independent.

5. BIBLIOGRAPHICAL REFERENCES

ABBUD, O. A; TRANCREDI, M. Recent transformations in the Brazilian electricity generation matrix - main causes and impacts. **Textos para Discussao**. v.1, n. 69, p.2-17, 2010.

ABNT - ASSOCIAÇÃO BRASILEIRA DE NORMAS TÉCNICAS. **NBR 6065: Determination of the Degree of Exhaust Emitted by Diesel Engine Vehicles or Equipment by the Free Acceleration Method**. Rio de Janeiro, 1980.

ABNT - BRAZILIAN ASSOCIATION OF TECHNICAL STANDARDS. **NBR 5422: Design of Overhead Electricity Transmission Lines**. Rio de Janeiro, 1985.

ABNT- ASSOCIAÇÃO BRASILEIRA DE NORMAS TÉCNICAS. **NBR 7027: Automotive Road Vehicles - Diesel Engine Emitted Smoke - Determination of Opacity or Degree of Blackening at Constant Regime**. Rio de Janeiro, 2001.

ABNT - ASSOCIAÇÃO BRASILEIRA DE NORMAS TÉCNICAS. **NBR 6016: Diesel Engine Exhaust Gas - Evaluation of Soot Content with the Ringelmann Scale**. Rio de Janeiro, 2015.

ABRADEE - BRAZILIAN ELECTRIC ENERGY AGENCY. Electricity **Sector Electricity Networks**. 2016. Available at:<http://www.abradee.com.br/setor-eletrico/redes-de-energia-eletrica>. Accessed on: October 12, 2016.

ALMEIDA, M. R. R.; ALVARENGA, M. I. N.; CESPEDES, J. G. **Quality analysis of Environmental Control Reports approved by the Regional Superintendence for the Environment and Sustainable Development of Southern Minas Gerais**. Dissertation (Master of Science in Environment and Water Resources). Federal University of Itajubà, Itajubâ. 5p. 2010.

ANEEL - NATIONAL ELECTRIC ENERGY AGENCY. **Atlas of Electricity in Brazil**, 1ª ed. Brasilia: ANNEL, 2002. 153p.

ANEEL - NATIONAL ELECTRIC ENERGY AGENCY. **Atlas of Electricity in Brazil**, 2nd ed. Brasilia: ANNEL, 2005. 20p.

ANEEL - NATIONAL ELECTRICITY AGENCY. **Atlas of electricity in Brazil**. 3a ed. Brasilia: ANEEL, 2008. 236p.

ANEEL - NATIONAL ELECTRICITY AGENCY. **Resolution No. 427, of February 22, 2011.** 1st ed. Brasilia: ANEEL, 2011. 29p.

ANEEL - AGÊNCIA NACIONAL DE ENERGIA ELÉTRICA, Anexo 6AB LOTE AB do edital de leilâo no 011/2013-ANEEL. Brasilia: ANEEL, 2013. 59p.

BELO MONTE. **The company.** 2016. Available at: <http://www.bmte.com.br/the-company/>. Accessed on: October 23, 2016.

BRAZIL, **Law. 6.938, of August 31, 1981.** Provides for the National Environmental Policy. Available at: <http://www.planalto.gov.br/ccivil_03/leis/L6938.htm>. Accessed on: October 22, 2016.

BRAZIL, **Resolution No. 001, of January 23, 1986.** Provides for basic criteria and general guidelines for environmental impact assessment. Available at:< http://www.mma.gov.br/port/conama/res/res86/res0186.html>. Accessed on: October 22, 2016.

BRAZIL, **Decree No. 97.632, of April 10, 1989.** Provides for the regulation of Article 2, item VIII, of Law No. 6.938, of August 31, 1981, and makes other provisions. Available at:< http://www.planalto.gov.br/ccivil_03/decreto/1980-1989/D97632.htm>. Accessed on: October 22, 2016.

BRAZIL, **Law No. 8.171, of January 17, 1991.** Provides for agricultural policy. Available at:<https://www.planalto.gov.br/ccivil_03/Leis/L8171.htm>. Accessed on: October 22, 2016.

BRAZIL, **Law No. 9.427, of December 26, 1996.** Establishes the National Electric Energy Agency (ANEEL), regulates the regime for public electricity service concessions. Available at:<http://www. planalto. gov. br/ccivil/Leis/L9427cons. htm>. Accessed on: October 22, 2016.

BRAZIL, **Resolution No. 237, of December 19, 1997.** Provides for and regulates the aspects of environmental licensing established in the National Environmental Policy. Available at:<http://www.mma.gov.br/port/conama/res/res97/res23797.html>. Accessed on: October 22, 2016.

BRAZIL, **LAW No. 9.608, of February 18, 1998.** Provides for voluntary service and other measures. Available at:<http://www.planalto.gov.br/ccivil_03/leis/L9608.htm>. Accessed on: October 22, 2016.

a. BRAZIL, **Law No. 10.847, of March 15, 2004**. Establishes the Energy Research Company - EPE. Available at:<http://www.planalto.gov.br/ccivil_03/_ato2004-2006/2004/lei/l10.847.htm>. Accessed on: October 22, 2016.

b. BRAZIL, **Law No. 10.848, of March 14, 2004**. Commercialization of electricity. Available at:<http://www.planalto.gov.br/ccivil_03/_ato2004-2006/2004/lei/l10.848.htm>. Accessed on: October 22, 2016.

c. BRAZIL, **Opinion No. 312/CONJUR/MMA/2004**. Provides for basic criteria and general guidelines for environmental impact assessment. Available at:<www.ibama.gov.br/licenciamento/modulos/arquivo.php?cod_arqweb=par312>. Accessed on: October 22, 2016.

a. BRAZIL, **Normative Instruction No. 4, of April 13, 2011**. Establishes procedures for preparing a Degraded Area Recovery Project - PRAD or Altered Area.

Available at:<http://www.ibama.gov.br/sophia/cnia/legislacao/IBAMA/IN0004-130411.PDF>. Accessed on: October 22, 2016.

b. BRAZIL, **Resolution No. 429, of February 28, 2011**. Provides on the methodology for recovering Permanent Preservation Areas - APPs. Available at:<http://www.mma.gov.br/port/conama/legiabre.cfm?codlegi=644>. Accessed on: October 22, 2016.

BRAZIL, **Law No. 18.104 of July 18, 2013**. Provides for the protection of native vegetation, institutes the new Forest Policy of the State of Goiás and makes other provisions. Available at:<https://www.legisweb.com.br/legislacao/?id=256749>. Accessed on: October 22, 2016.

CAMPOS, O. L. Case study on the environmental impacts of transmission lines in the Amazon region. **BNDES Setorial, Rio de Janeiro,** n. 32, 2010. p. 231-266.

CORRÊA, A. N. S. **Beef cattle: the producer asks, Embrapa answers.** Embrapa Publishing House, 1996. V.0, 208 p.

ELETROBRAS - CENTRAIS ELÉTRICAS BRASILEIRAS S. A. **How electricity is transmitted in Brazil.** 2015. Available at:

<http://www.eletrobras.com/elb/natrilhadaenergia/energia-

eletrica/main.asp?View=%7B05778C21-A140-415D-A91F-1757B393FF92%7D>.
Accessed on: October 16, 2015.

EPE - EMPRESA DE PESQUISA ENERGÉTICA, **Anuàrio Estatistico de Energia Elétrica 2013,** 1ª ed. Rio de Janeiro: EPE, 2014. 253p.

a. EPE - EMPRESA DE PESQUISA ENERGÉTICA, **Anuàrio Estatistico de Energia Elétrica 2015**, 1st ed. Rio de Janeiro: EPE, 2015. 232p.

b. EPE - EMPRESA DE PESQUISA ENERGÉTICA, **Electricity Demand Projections 2015-2014,** 1st ed. Rio de Janeiro: EPE, 2015. 90p.

EPE - EMPRESA DE PESQUISA ENERGÈTICA, **Balanço Energético Nacional 2016: Relatório Sintese - ano base 2015**, la ed. Rio de Janeiro: EPE, 2016. 62p.

FERREIRA, A. P. B. da S. **Problematics and prospects for the use of permanent magnet generators in wind energy production**. Dissertation (Specialization in Energy Systems). University of Porto. Porto. 230p. 2012.

FERREIRA, J. B. **Study of environmental impacts and mitigating measures for an electricity transmission line project**. Course Conclusion Paper (Agronomic Engineering). Federal University of Santa Catarina. Florianópolis. 48p. 2011.

FERREIRA, R. D. E. **Methodology for the efficient application of solar energy in homes**. Course Conclusion Paper (Electrical Engineering). Federal University of Rio de Janeiro, Rio de Janeiro. 140p. 2016.

GRUPO COBRA, **Presentation of the Company**. 2016. Available at: <http://www.grupocobra.com/content/page/presentation-of-the-company>. Accessed on: October 23, 2016.

MENEZES, V. P. **Electric power transmission lines: technical, budgetary and construction aspects**. Course Conclusion Paper (Electrical Engineering). Federal University of Rio de Janeiro, Rio de Janeiro. 75p. 2015.

MORENO, R. F.; MORENO, L. C. R. Possible effects on human health resulting from exposure to low frequency electric and magnetic fields - commented review of the literature. In: XVI SNPTEE - Seminàrio Nacional de Produçâo e Transmissâo de Energia

Elétrica, 8., 2001, Campinas. **Anais...**, Campinas: GIA/026, 2001. p.1-6.

ONS - NATIONAL ELECTRICITY SYSTEM OPERATOR. **Maps of the national interconnected system**. Rio de Janeiro: ONS, 2015. Available at: <http://www.ons.org.br/conheca_sistema/mapas_sin.aspx>. Accessed on: October 16, 2015.

OTTMANN, M. M. A.; BORCIONI, E.; MIELKE, E.; CRUZ, M. R.; FONTE, N. N. Environmental and socio-economic impacts of community gardens under transmission lines in the Tatuquara neighborhood, Curitiba, PR, Brazil. **Revista Brasileira de Agroecologia**, v. 5, n. 1, 2010.

PBA - BASIC ENVIRONMENTAL PROJECT - **LT CC ±800 _kV Xingu / Estreito and Associated Facilities: P.03 - Degraded Areas Recovery Program (PRAD)**. 2015.

Available at: <http://licenciamento.ibama.gov.br/Linha%20de%20Transmissao/LT-800-kV-XinguEstreito/PBA/P.03%20-%20PRAD%20rev2.pdf>. Accessed on: October 22, 2016.

PEREIRA, E. A.; MONTARDO, D. P.; DECKMANN, P.; DHEIN, R. Adaptability and stability of selected sudan grass populations in different environments in rio grande do sul. In: XVII Congress of Scientific Initiation. X Encontro de Pos-Graduaçâo, 7., 2008, Pelotas. Proceedings..., Pelotas: UFP, 2008. p.1-4.

PIRES, S. H. M. **Desafios ambientais no novo modelo do setor elétrico**. 1ª ed. Rio de Janeiro: FBDS, 2005.

REIS, A., TRES D. R.; SIMINSKI, A. **Recovering degraded areas: Imitating nature**. 1st ed. Florianópolis: Rio Negrinho, 2005. 90p.

ROMANO, R. L. **Robust adaptive dual-mode control applied to an electric power generation system**. Dissertation (Master's Degree in Automation and Systems). Federal University of Rio Grande do Norte, Natal. 57p. 2015.

SÂ, V. S. **Feasibility study on the use of grid-connected photovoltaic generation systems in Brazil**. Course Conclusion Paper (Civil Engineering). Federal University of Ouro Preto, Ouro Preto. 40p. 2016.

SANTOS, M. dos. **Plan for the recovery of degraded areas along the electricity transmission line in the municipality of Passos Maia-SC**. Course Conclusion Paper

(Agronomic Engineering). Federal University of Santa Catarina, Florianópolis. 65p. 2011.

SANTOS, R. H. **Integrated Resource Planning and Regulation: the US experience and prospects in Brazil**. Dissertation (Master's Degree in Energy). University of Sâo Paulo, Sâo Paulo. 65p. 1997.

TOYAMA, A. H. JUNIOR, N. N.; ALMEIDA, N. G. **Economic feasibility study of the implementation of photovoltaic systems connected to the electricity grid for different regions in the state of Paranâ**. Course Conclusion Paper (Agronomic Engineering). Federal Technological University of Paranâ, Curitiba. 113p. 2014.

TRIGUEIRO, A. **Meio ambiente no século 21: 21 especialistas falam da questão ambiental nas suas áreas de conhecimento**. 2nd ed. Editora Sextante, 2005. 367p.

I want morebooks!

Buy your books fast and straightforward online - at one of world's fastest growing online book stores! Environmentally sound due to Print-on-Demand technologies.

Buy your books online at
www.morebooks.shop

Kaufen Sie Ihre Bücher schnell und unkompliziert online – auf einer der am schnellsten wachsenden Buchhandelsplattformen weltweit! Dank Print-On-Demand umwelt- und ressourcenschonend produziert.

Bücher schneller online kaufen
www.morebooks.shop

info@omniscriptum.com
www.omniscriptum.com

Printed by Books on Demand GmbH, Norderstedt / Germany